Birthday Book Club

This book was donated

in honor of the birthday of

Cole Nye-Smith

by

Mom and Dad

April 19, 1998

Aska's Sea Creatures

Other books with paintings by Warabé Aska

Seasons,
with poetry selected by Alberto Manguel

Aska's Animals,
with poetry by David Day

Aska's Birds,
with poetry by David Day

Aska's Sea Creatures

Paintings by Warabé Aska   Poetry by David Day

A Doubleday Book for Young Readers

A Doubleday Book for Young Readers

Published in Canada by Doubleday Canada Limited
105 Bond Street, Toronto, Ontario M5B 1Y3

Published in the United States by Delacorte Press
Bantam Doubleday Dell Publishing Group, Inc.,
1540 Broadway, New York, New York 10036

Art copyright © 1994 by Warabé Aska (Takeshi Masuda)
Text copyright © 1994 by David Day

Canadian Cataloguing in Publication Data
Aska, Warabé
    Aska's sea creatures

ISBN 0-385-32107-4

1. Aska, Warabé.   2. Aquatic animals - Pictorial
works. 3. Aquatic animals - Juvenile poetry.
I. Day, David, 1947-   . II. Title

ND249.A84A4 1994   j759.11   C94-930278-3

Library of Congress Cataloging in Publication data applied for

Cover design by Tania Craan
Printed and bound in Hong Kong
1994 Canada/1995 USA

This is a game you can play just for fun,
As you swim with the sea creatures chasing the sun.
Pretend you're a turtle, a fish or a whale
Pretend you have scales, or fins, or a tail....

Humpback Whale

Titanic opera singer, a diva of the deep,
Your voice resounds through the fishy world.
Your notes reach down into the depths and
  up again to the heavens.

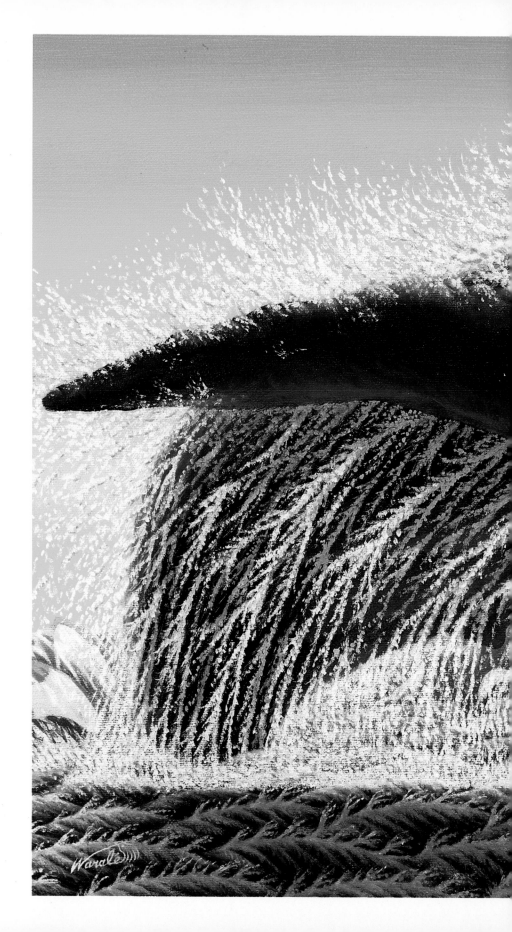

Dolphin

You feel playful and clever in the
   dolphin school.
Leaping, laughing and chattering,
You ride a wave of happiness
Over the wide sea of sorrows.

Great White Shark

You feel so hungry you could swallow
  the sun.
With chain saw teeth and sandpaper skin,
You are huge and terrible as a tidal wave.

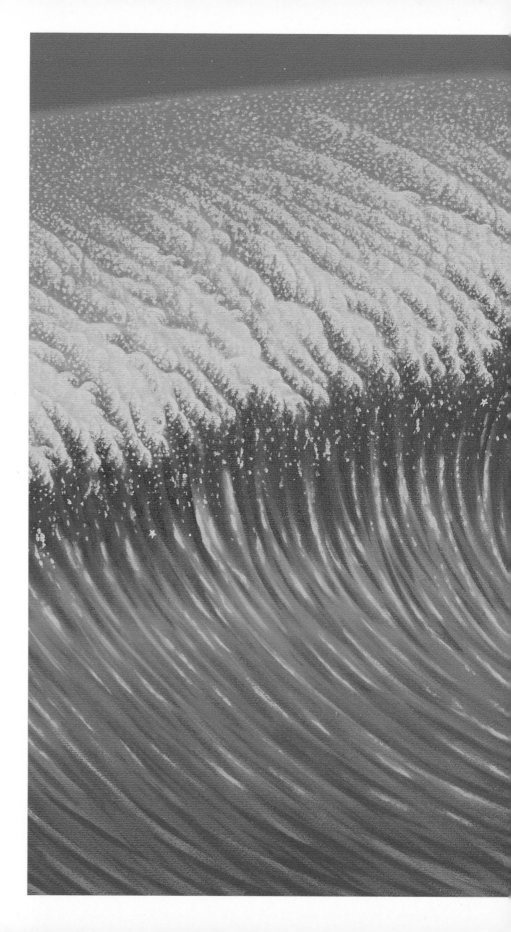

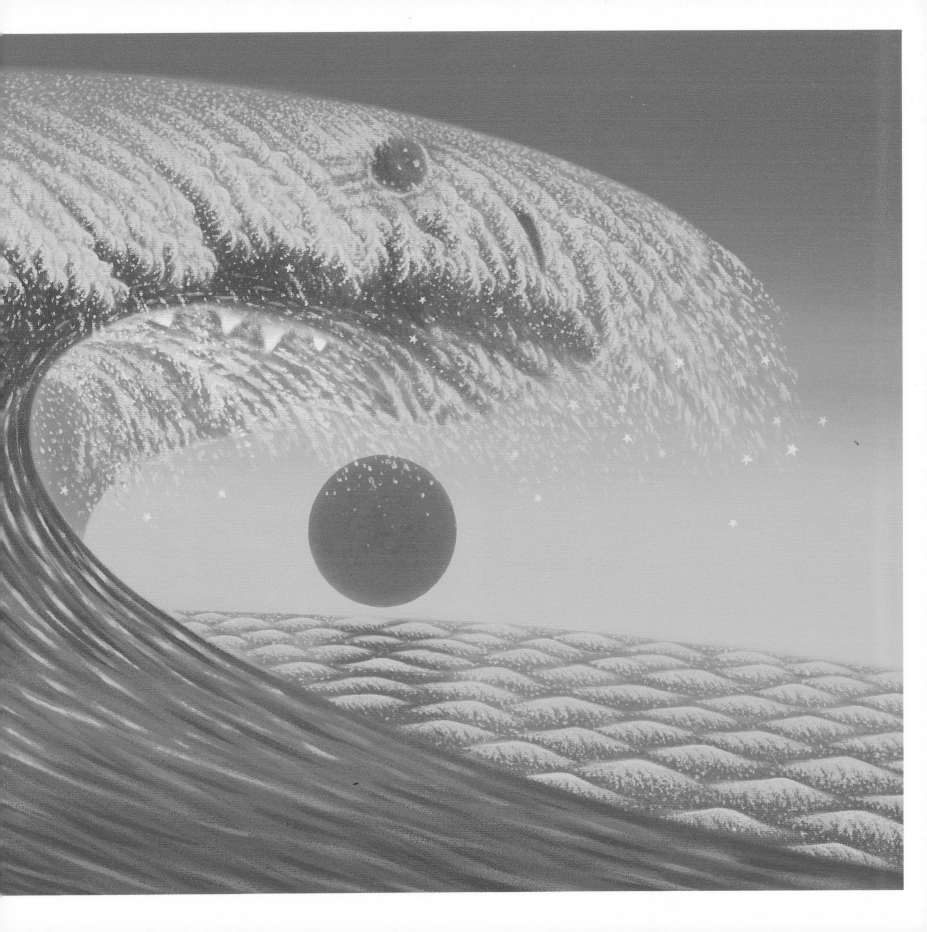

Starfish

You are tiny and you shine beneath the
   turquoise sea.
Your watery galaxy is a small mirror
Held up to the stars glinting above.

Puffer Fish

Comical and blown up like a spiky
  balloon,
You look like a prickly pear
And feel just like a clown !

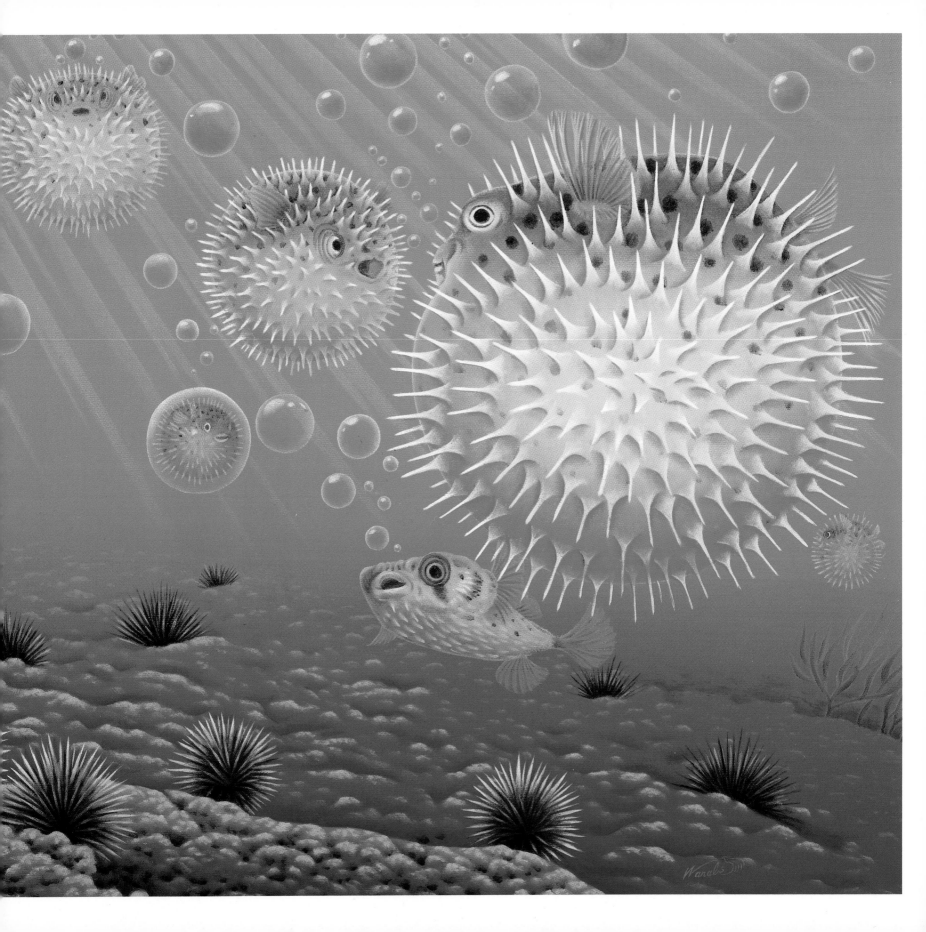

Seaweed

Waving hello, waving good-bye,
If you stay in one place long enough
All the world will drift past your door.

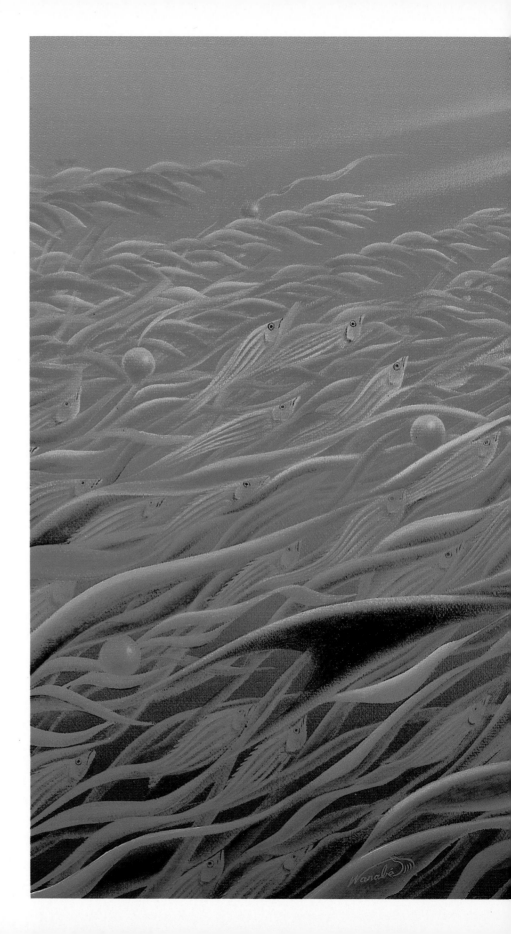

Sea Horse

You feel as free as a young colt
Let loose in the wilderness of the rocky
  reef,
Playing endless games of hide and seek.

Octopus

The ocean floor belongs to you,
Flinging your arms far and wide,
Any creature who dares come too close
Will be in for an inky surprise!

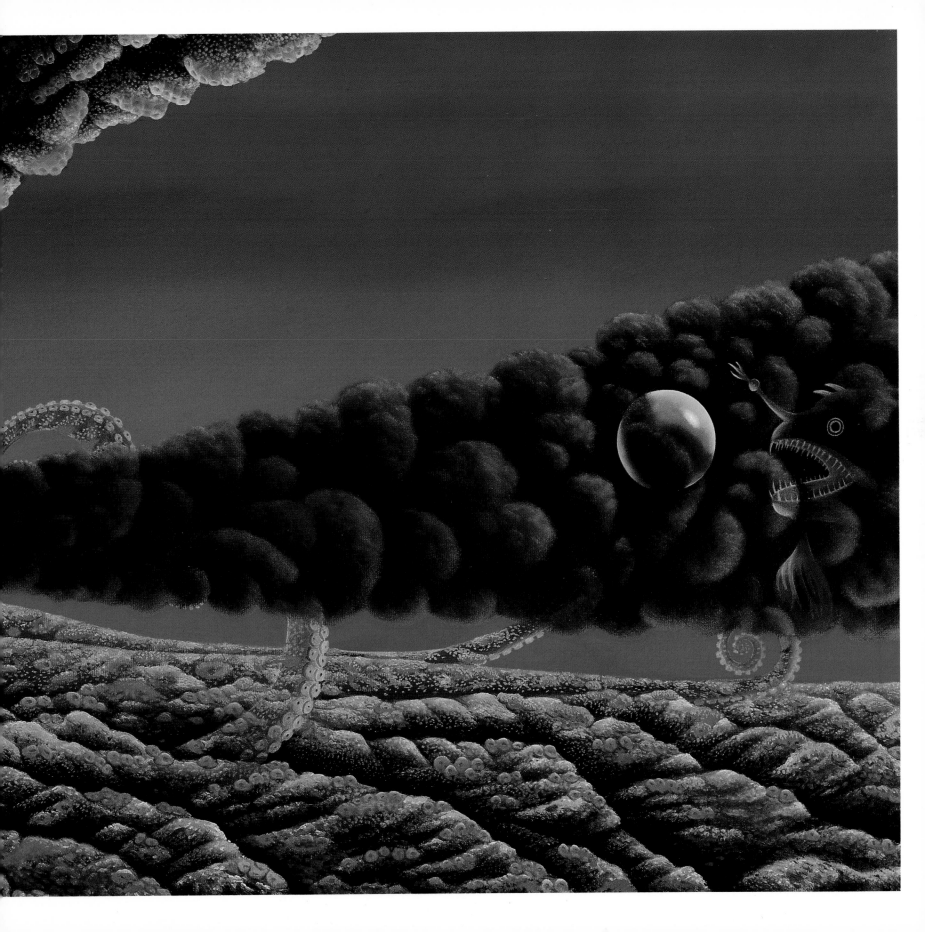

Stingray

You fly like a bat beneath the sea,
Or a jet no radar can track in flight.
Swooping down silently,
You nestle on the velvety sand.

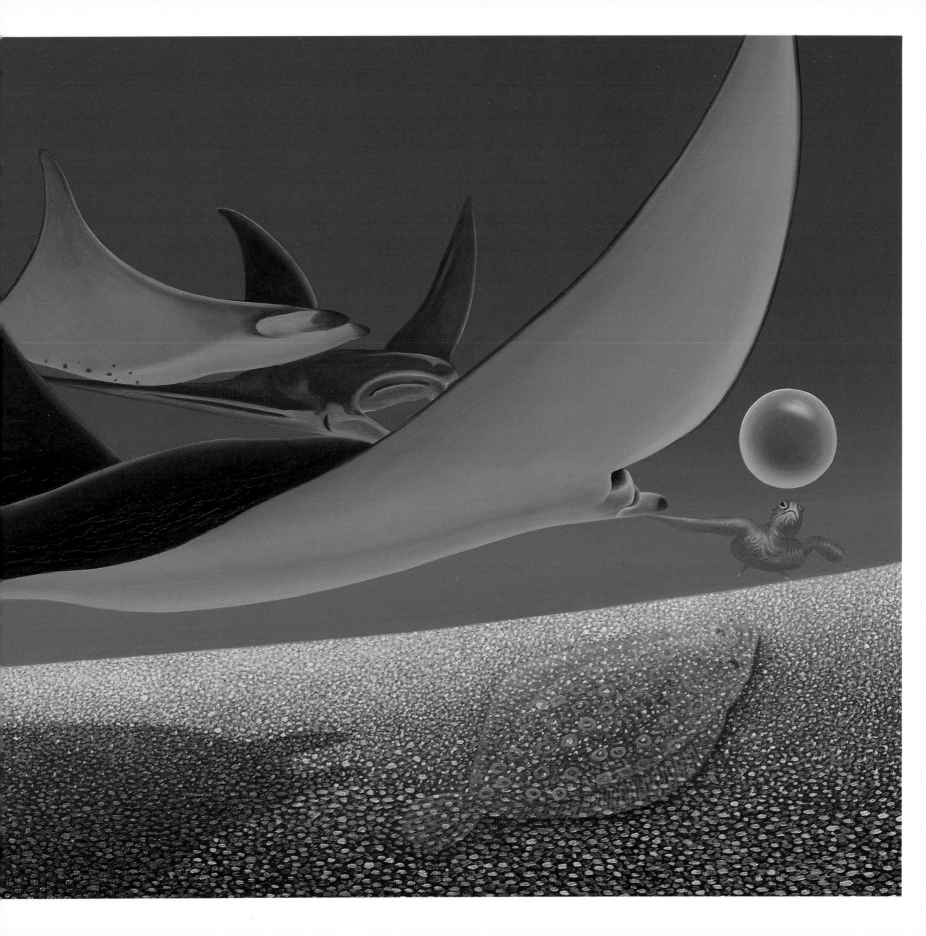

Angler Fish

No need for a mask at Halloween,
You're a monster fisherman of the deep
Luring small prey to your murky lair.

Sea Turtle

At your best when all alone at sea,
A solo sailor in your houseboat shell,
You carry with you all you need.

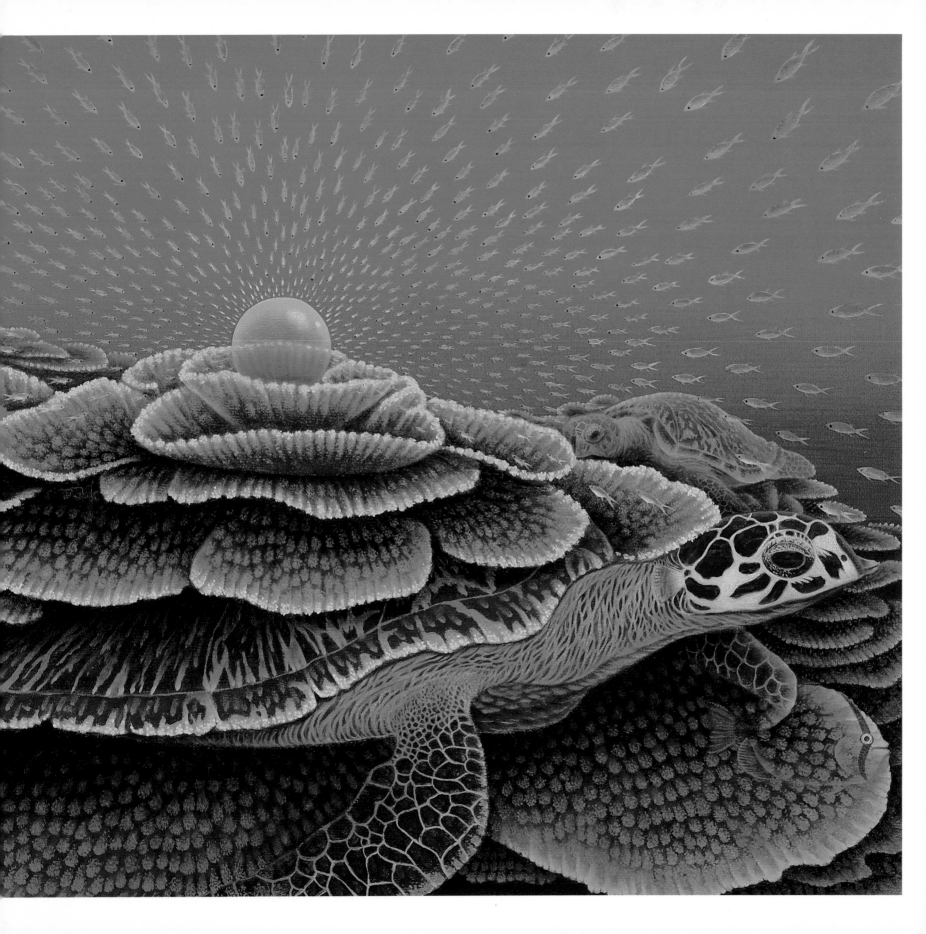

Coral

Fish gangs riot in your brilliant
   fairground,
A scary adventure playground
Where bully fish lurk.

Butterfly Fish

Happy and carefree,
You blow bubbles joyfully,
Fluttering gold fins in a tropical sea.

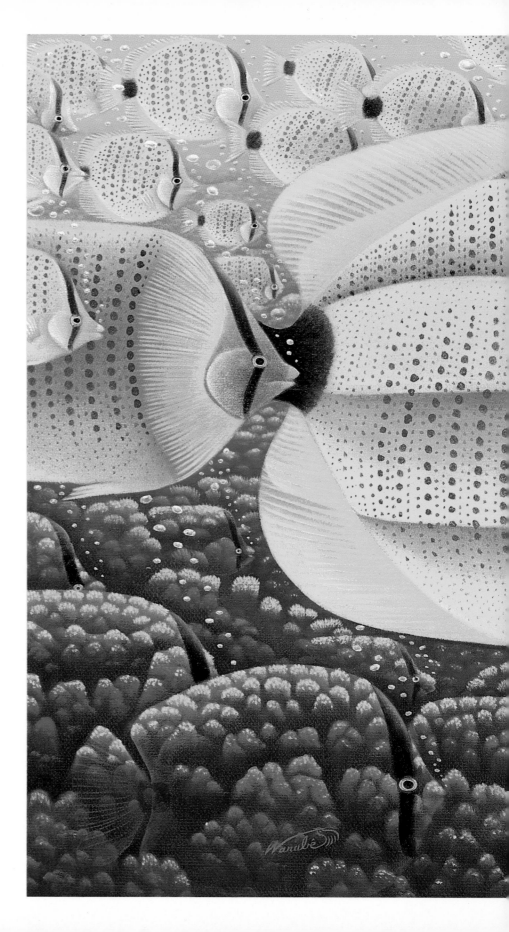

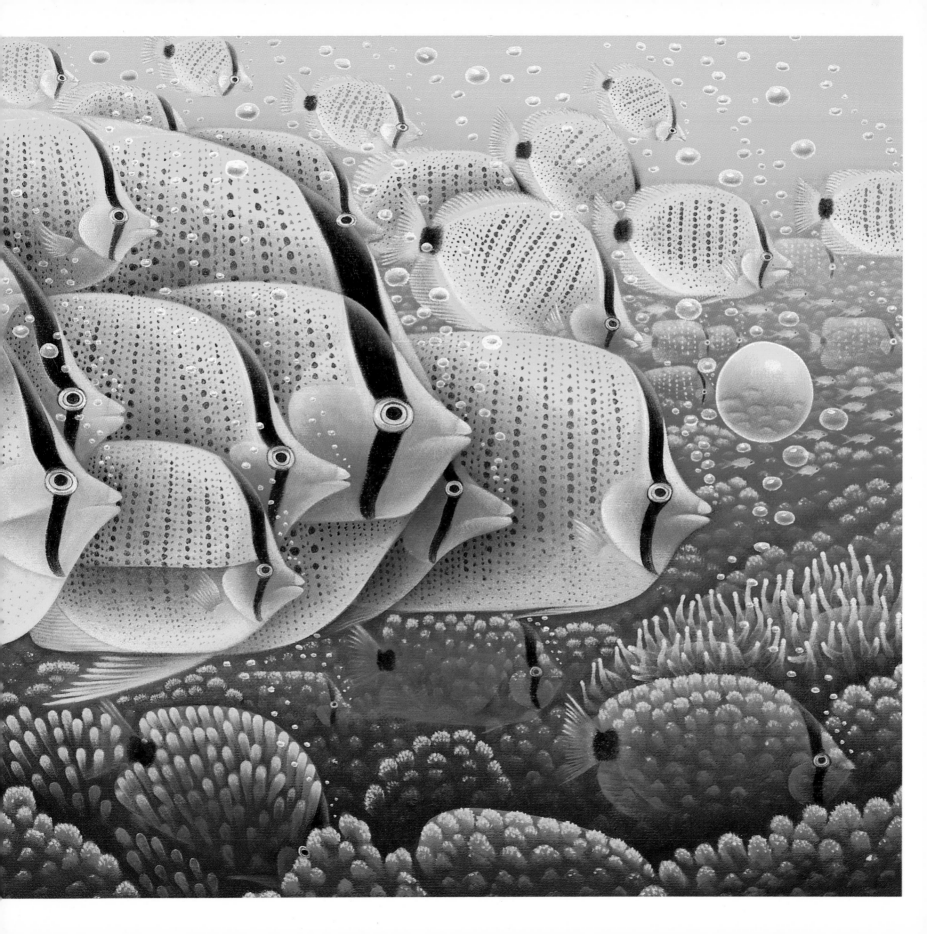

The day has come full circle around,
Time for sea creatures to sleep.
Time for the sun to go down.

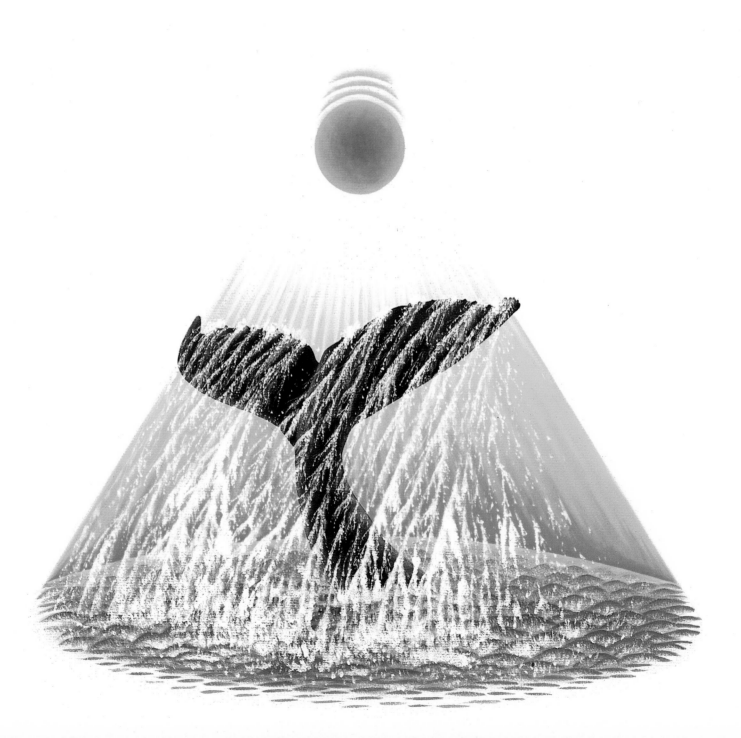